SPEEDSTERS

SPEEDSTERS

TODAY'S AIR RACERS IN ACTION

Philip Handleman

Airlife
England

Dedication

To the memory of my mother and father, tailwinds in Heaven

The Author

Philip Handleman has enjoyed a lifelong love affair with aviation. Since his first flight in a Piper Cub at 12 years of age, he has steadfastly pursued his interest in aeroplanes and flying. He has been a licensed pilot for more than 25 years, and currently owns and flies two restored aircraft of military lineage, including an open-cockpit Stearman biplane.

The author of nine previous aviation books and many dozens of aviation articles, Handleman is a recognised aviation authority. For many years he has beautifully captured a wide range of aeroplanes in his still photography. In addition, he is president of Handleman Filmworks, an award-winning television production company that has produced highly acclaimed documentaries. He and his wife, Mary, divide their time between their home in Birmingham, Michigan and a private airstrip in the nearby countryside.

Author's Note

But for a few scenes in which pilots and crews are posed with their aeroplanes, all of the photographs in this book are candid views. While there is certainly a market for prearranged shots of lovely and interesting aeroplanes, the photography contained herein, with the exceptions noted above, offers an unstaged, documentary perspective on modern air racing.

Speeds given for closed-course pylon races are generally the average speeds attained by the racing aeroplanes. Because an aeroplane blasting around such a course may go faster along the straightaway, or may record a better time on one lap than on another, it is important to realise that the overall performance is reflected in the average speed. In this book, unless it is indicated otherwise, the speeds for closed-course pylon races are average speeds.

First published in the UK in 1996
by Airlife Publishing Ltd

British Library Cataloguing in Publication Data
A catalogue record for this book
is available from the British Library

ISBN 1 85310 745 X

Typeset by Livesey Ltd, Shrewsbury
Printed in Hong Kong

Airlife Publishing Ltd

101 Longden Road, Shrewsbury SY3 9EB, England

Acknowledgements

Without the cheerful co-operation of personnel at the official air racing organisations, the Reno Air Racing Association, Inc. and the Superstition Racing Corporation, this project simply would have been unfeasible. Media relations specialists Zach Spencer and Dave Lenzen, of the Reno and Phoenix air races, respectively, provided invaluable assistance for which I am deeply grateful. At Reno, the Washoe Jeep Squadron, a ragtag but highly motivated band of search and rescue workers, deserves thanks for supplying a sometimes acerbic press corps with a reliable shuttle service to the pylons.

My wife, Mary, contributed the most to this endeavour. When backaches threatened to interrupt my photography, she unhesitatingly bent over, picked up a load of camera gear, and led the way. Later, when other forms of interference weighed in, she lent comfort and encouragement. Mary is my navigator in life as in flight.

PHOTOGRAPHS:

JACKET FRONT, PAGE 2 AND TITLE:
A dogfight between Pitts biplanes. In the lead is *Ole Yeller*, flown by retired airline pilot Ray Kraspovich of Arvada, Colorado, with *My Pitts*, piloted by South Africa Airlines captain Levin Scully, closing fast. Captain Scully, a first-time participant at Reno in 1995, did surprisingly well.
The winner of the Biplane Class Race at Reno in 1995, with a speed of 202.124 mph, was Patti Johnson, an aerobatic flight instructor from Edgewater, Florida. In 1993, piloting her Mong, named *Full Tilt Boogie*, she became the first woman to win a racing-class championship. In so doing she has admirably built upon the traditions of predecessors such as Laura Ingalls, Louise Thaden, Elinor Smith, Pancho Barnes, Jacqueline Cochran, and Amelia Earhart. *(Reno, 1995)*

JACKET BACK:
Top Left:
Aerodynamic streamlining is evident in the Smith Mini biplane *Stinger* piloted by Charlie Chambers of Bend, Oregon. *(Reno, 1995)*
Bottom Left:
The Canadian Forces 'Snowbirds', officially known as 431 Air Demonstration Squadron. *(Reno, 1995)*
Top Right:
'Tiger' Destefani tore up the sky with *Strega*, reclaiming the Unlimited Gold crown at Reno in 1995 with a speed of 467.029 mph, the third fastest winning time in Reno air racing history. The severe angle of bank is necessary to keep the aircraft from drifting on to a wider pattern around the course. *(Reno, 1994)*
Bottom Right:
One of the many Pitts Specials in the Biplane Class, *Ole Eight Lima* belongs to Jeffrey Lo of San Jose, California. *(Reno, 1995)*

HALF TITLE:
Ironically, the two-tone tan camouflage scheme makes *Cottonmouth* more easily recognised on the race circuit. At Reno in 1994, airline pilot Bramwell Arnold of Lemoore, California, flew this old fighter to first place in the Bronze race at a speed of 301.401 mph. *(Reno, 1994)*

INTRODUCTION:
Helicopter pilot Thomas Hauptman rounding the pylons in the Wagner Shoestring *Judy*. *(Reno, 1995)*

Introduction

Air racing has been an integral part of my life for as long as I can remember. Among my first recollections of growing up were stories of an enchanted place where dashing airmen sped around a pylon-marked course in marvellously-adorned biplanes and scrupulously-streamlined monoplanes to the cheers of admiring spectators. What made these stories so special is that they were genuine, being imparted to me first-hand by an attendee.

By a quirk of fate, I suppose, my mother's family settled near Cleveland, Ohio, in the quaint village of Berea. Grandpa Harry, an extraordinarily dignified man, was the town barber and Grandma Anna, a stalwart of traditional values, remained preoccupied managing the large and ever growing household. Among the children, my mother was the lone girl. It was natural, then, that she not only developed into a tomboy but revelled in her hoydenish adolescence.

With the approach of Labor Day, when the first inkling of a colour change registers in the foliage and a wave of brisk autumn air passes through, Mother delighted in knowing that the air races were coming. On race days, she and like-minded neighbourhood kids strolled to the nearby airport and climbed over the fence. There, they beheld history's greatest aviation spectacle.

Until she died, the image of her favourite air racing pilot, the swashbuckling Roscoe Turner, was fresh in her memory. So many nights she told me of the wondrous times she had had at the air races. It was like a fairy tale that came true.

When she grew up, she went to work at that airport. She met my father there; they got married and moved away, but reminiscences of Cleveland always brought a warm glow to their eyes.

Today, when I stand on the ramp at Reno-Stead or at Willie and see youngsters gleefully focused on air racers turning with breakneck speed around the pylons, I suspect that in a generation they, too, will be passing on to their children the excitement, the fun, the glory of aeroplanes and pilots at their best.

Contents

If one were seeking an air race location, Reno would be a natural choice. A mile above sea level, blessed with crisp desert air, and away from crowded population centres, it makes a good setting for pylon style air racing. Brightly decorated pylons mark the course. *(Reno, 1995)*

Every year at the Reno air races, the dying art of skywriting gets a reprieve when Steve Oliver and his wife, Suzanne Ashbury-Oliver, fill the vivid morning sky with greetings to the race fans, using a vintage Travel Air biplane. A trail of biodegradeable smoke is emitted from the engine's exhaust, creating precise and elegant script in the three dimensions of the skywriter. The task is made to look simple by these superb pilots, who are sponsored by the maker of a popular soft drink. The company's name often accompanies the skywriters' aerial messages. *(Reno, 1994)*

Chapter 1
Air Racing Yesterday and Today

Surely, air racing began as soon as two competitively minded pilots flew separate aeroplanes within view of each other. But it was not until six years after the Wright brothers succeeded in flying by means of a powered and controllable heavier-than-air vehicle that organised air racing began.

Sanctioned by the aero clubs of France, England, and the United States, this first official air racing event occurred outside Paris, in the countryside near Reims. It was described as *La Grande Semaine d'Aviation* (The Great Week of Aviation), and enormous crowds gathered each day of the week-long event in August 1909.

Although the race grounds became increasingly strewn with the wrecks of fragile wood-and-fabric aircraft, the spirit of the pilots was not dampened. On the final day, Glenn Curtiss, the only American entered, won the coveted James Gordon Bennett Cup piloting his 'Reims machine' at an average speed of 47.65 mph. He bested famed Frenchman Louis Blériot, who came in second, by a mere 5.4 seconds. Organised air racing, primitive though it was, had been born.

Aviation, still in its infancy, offered a means to revolutionise transport, commerce, and warfare. Clearly, advances in the new science of aeronautics were necessary in order to realise the aeroplane's full potential. Cognizant of this fact, certain wealthy individuals, acting as patrons, sponsored air racing awards as a means to further aviation technology. By fostering competition, it was believed that progress would result.

The Bennett Cup Race became an annual event, the venue being shared among the countries of the three sponsoring aero clubs. Speeds steadily rose and, at Chicago in 1912, Jules Vedrines, flying a dramatically streamlined Deperdussin monoplane, achieved an average speed of over 100 mph.

The Bennett Cup Race was cancelled during the war years, and when it resumed in 1920 at Estampes, France, the impact of wartime advances was evident. Aircraft had progressed remarkably as a result of war-driven urgency. At the war's end, the French had emerged as a world aviation leader, and they were represented at the races by their latest military fighter aircraft.

The American aircraft industry had fallen behind and did not have any viable off-the-shelf contenders. However, the USA did have pockets of innovation, and that year's American entries reflected fresh design thinking. In subsequent years, the ability to innovate has been a key ingredient in air racing success.

That year, the Army Air Service's entry, a Verville VCP-1 Scout, enjoyed a significant power advantage, its normal production engine being replaced by a 638 hp Packard V-12. Also, Curtiss built a racing aeroplane on commission for an oil tycoon. Called 'Texas Wildcat,' this aircraft featured a unique enclosed cockpit, set back near the tail.

The most futuristic entry was the Dayton-Wright RB-1 racer, the only monoplane in the race. Its wing was fully cantilevered, the inner panels consisting of a gridwork of balsawood. The wing's camber could be varied by the use of hinged surfaces on both the leading and trailing edges. This purpose-built machine also benefited from a retractable undercarriage, a first for air racers. The monocoque-constructed fuselage was quite narrow to minimise drag, and the pilot had no direct forward view; he compensated by using a movable back-and-forth side window.

Although the Dayton-Wright hybrid was clocked at 200 mph, it suffered a broken rudder cable which forced it out of the race. Similarly, the Curtiss racer achieved a speed of 215 mph, but crashed before the race. Even the promising Verville Scout with its souped-up powerplant had to drop out because of mechanical problems.

The winner of the race was a Nieuport, which recorded an average speed of 168.5 mph. This, the last of the Bennett Cup Races, graphically illustrated that speed by itself does not guarantee victory. It became clear that improved designs and innovative modifications may enhance chances of success, but other factors, such as structural sturdiness and flying skill are critical too.

Recognising the potential in expanding aircraft versatility to include water operations, wealthy Frenchman Jacques Schneider established an international seaplane competition a few years after the first air meet at Reims. The first Schneider Trophy Contest was held at the seaside resort of Monaco in April 1913, and the winning speed was a mere 45.75 mph. One more Schneider Trophy Contest occurred before the war interrupted what was on the way to becoming an annual event. The lessons learned from these races were turned to practical use, the military aircraft designs of the First World War being influenced by the strides made in air racing. This ominous phenomenon was to be repeated later.

The 1920s saw a renewed series of Schneider Trophy Contests featuring colourful seaplanes, from the USA, France, England, and Italy. Winning speeds steadily increased over the years, with beautiful aircraft festooned with bulging pontoons such as the Macchi M.39 or the Curtiss R3C-2 to the fore.

By 1931, England was leading seaplane design with the Supermarine S.6B, powered by a Rolls-Royce 'R' engine, which won the trophy outright for Great Britain. At Lee-on-Solent, Lieutenant J. H. Boothman pushed his racer to a new course record of 340.1 mph. This seaplane formed the basis for the legendary Spitfire, which played a crucial role in the Battle of Britain. Also, the racer's engine evolved into the superb Merlin, which became the Spitfire's powerplant.

Two years after the Schneider Trophy was claimed by Great Britain, the world's seaplane (and landplane) speed record was broken by the Macchi-Castoldi MC.72. This ingeniously designed seaplane had two engines installed in tandem in an elongated nose, driving a pair of contra-rotating propellers. The MC.72 attained the astounding speed of just over 440 mph on a straight 3 km course. Although a landplane finally exceeded this speed in 1939, the seaplane record stands to this day.

In the aftermath of the First World War it was apparent even to those who had previously been sceptical that the aeroplane was not a mere technological curiosity but an invention of far-reaching import. Possessing the lead in this burgeoning field became an important national objective.

Ralph Pulitzer, a scion of the famous American publishing family, prompted a new annual speed contest bearing his family's name. The Pulitzer Trophy Races, introduced in 1920, were open to competitors from any country. They brought the excitement of air racing to the heartland of America, being held in cities scattered across the nation – Omaha, Detroit, St Louis (birthplace of the Pulitzer fortune), Dayton, and on Long Island, New York.

A trend emerged with Bert Acosta's 1921 win in a Curtiss R-1. All but two of the six Pulitzer Trophy Races were won by Curtiss-designed aeroplanes. Their sleek fuselages and powerful 12-cylinder, liquid-cooled engines presaged the famous line of Curtiss Hawk pursuit aircraft.

At the last race, in 1925, winning speeds topped out at nearly 250 mph, a new record for closed-course racing. The USA military, which had funded the specially conceived Curtiss racers, then decided to conclude its support. It was the end of the Pulitzer Trophy Races.

The National Air Races began the following year, when the National Aeronautic Association, the official aero club of the United States, combined smaller races scattered around the country into a single event. Philadelphia was chosen as the venue, and the main prize was a Kansas City Rotary Club Trophy. The award went to a naval pilot flying the new Boeing FB-3. As this was a production fighter, not a purpose-built racer, its winning speed was just 180.495 mph, substantially lower than the previous year's winning speed at the Pulitzer Trophy Race.

Several years earlier a tradition was started in which the Army's First Pursuit Group pilots competed among themselves in a race flying identical pursuit aeroplanes. Named in memory of John L. Mitchell, the deceased brother of outspoken airpower advocate William 'Billy' Mitchell, this race was always evenly matched and thrilling to watch, though the speeds were not record-setting. The Mitchell Trophy Races occurred in concert with the larger air racing events.

The 1927 National Air Races were held in Spokane, Washington. Predictably the military dominated, a modified Curtiss Hawk XP-6A taking first place at an average speed of 201.239 mph. The highest award was the *Spokane Spokesman-Review* Trophy.

The next year, a military-free-for-all was staged in what was to become the mecca for aviation, Los Angeles. At Mines Field (now Los Angeles International Airport) a Navy XF4B-1, a new Boeing designed fighter, won the military race at an average speed of only 172.26 mph. Civilians raced separately that year, and a brand-new Lockheed Vega, a production aircraft, won this race and captured the *Detroit News* Trophy. During the air show part of the event, fantastic flying displays were put on by Army and Navy formation teams named the Three Musketeers and the Three Sea Hawks, respectively.

In 1929, the National Air Races took place in Cleveland, Ohio, for the first time. This successful event was a harbinger of things to come. Hundreds of thousands came to see the races, and the civilian entries, unlike those of the immediately preceding years, were genuinely competitive with the military's finest pursuit aircraft in terms of performance. A snappy low-wing monoplane from the Travel Air Company of Wichita, Kansas, the Model R 'Mystery', crossed the finishing line first with a comfortable lead over an Army Curtiss P-3A Hawk.

Entrepreneurial inventors were seizing the initiative and producing aeroplanes with better performance than those being procured by the War and Navy Departments. This fact was reinforced at the 1930 National Air Races in Chicago, where racing legend Charles 'Speed' Holman came in first, flying the Laird 'Solution' biplane at an average speed of over 200 mph. In doing so, he won the first Thompson Trophy, sponsored by Thompson Products (now TRW, Inc.).

In 1931, the National Air Races were moved back to Cleveland, where, except for two years in Los Angeles, they stayed for the remainder of the decade. Modern air racing can trace its origins to Cleveland. Increasingly, one-of-a-kind aeroplanes, specially built or modified, would outrun the fighters of the day. The Thompson Trophy became a symbol of aeronautical prowess, and the racing aircraft and their intrepid pilots captivated the public.

That same year, industrialist Vincent Bendix offered a trophy for the winner of a cross-country contest. Most of these races were dashes from Burbank, California, to Cleveland. The first recipient of the Bendix Trophy was none other than the outstanding Jimmy Doolittle, flying the Laird 'Super Solution.' The achievements of these aerial daredevils offered a glimmer of hope, a prospect of better times, to a downtrodden nation suffering the throes of the Great Depression.

Unlike the closed-course Thompson Trophy Races, the cross-country Bendix Trophy Races were open to women. Louise Thaden won in 1936, flying a Beechcraft C-17R Staggerwing and, two years later, Jacqueline Cochran was victorious in a Seversky SEV-S2. These great women flyers and others like them were not only winning races and setting speed records, but were bringing sexual equality closer to realization through their achievements.

The previous closed-course record set in the 1925 Pulitzer Trophy Race fell in 1932 to Jimmy Doolittle, who piloted the barrel-like Gee Bee Senior Sportster R-1. He captured the Thompson Trophy at a speed of 252.686 mph.

Another category of racer was recognised with the establishment in 1934 of the Greve Trophy Races for aircraft powered by relatively small engines not exceeding 550 in^3 of displacement. A group of racing pilots including notables such as Harold Neumann, Steve Wittman, and Art Chester excelled at flying these 'midgets', powered for the most part by Menasco engines. The highest speed attained in this class was 263.390 mph in 1939, when Art Chester's 'Goon' outlasted the other contenders at Cleveland.

Each Labor Day weekend the nation's attention was focused on the Thompson Trophy Race, for it was the grandaddy of pylon races. Of the great pilots participating in these races and the accompanying air shows, none so enthralled the public as the inimitable Roscoe Turner.

Born to an impoverished farm family in rural Mississippi, Turner was blessed with a mechanical bent from childhood. After serving as an Army airman in the closing days of the First World War, he became a barnstorming pilot, criss-crossing the country selling aeroplane rides and putting on flight displays. His flying abilities were honed as he accumulated many hours in rickety old aircraft, and aviation became second nature to him.

His flying skills were matched with a consummate showmanship. At the air races, Roscoe appeared in a custom tailored military-style tunic replete with Sam Browne belt, jodhpurs, and polished boots. He sported a waxed mustache that added character to his chiselled features and broad smile.

Fred Crawford, an officer of Thompson Products during the 1930s and destined to become president of the aircraft parts company, recalled that at the time Roscoe was often in debt. According to Crawford, at one of the first races in Cleveland a process server turned up at the airfield, intending to repossess the aspiring champion's aeroplane. A clever mechanic kept the man at bay long enough for Roscoe to jump into his racer and compete in the race. He finished first, and used the prize money to stave off the impending loss of his aeroplane.

For publicity, Roscoe appeared with a pet lion. The newsreel cameras caught his dashing profile; he was a larger-than-life character, and it was almost as though air racing had been created for his participation. Central casting could do no better than the real life Roscoe Turner. Roscoe was the only person to win the Thompson Trophy Race three times. In 1938 he established the pre-war pylon racing record with an average speed of 283.419 mph.

In 1939, the last pre-war year of the Thompson Trophy Races, a new speed record was set on the other side of the Atlantic Ocean, when German Fritz Wendel flew the Messerschmitt Me-209V-1 over a straight 3 km course at a sizzling 469.22 mph. German propagandists falsely spread the notion that this one-off prototype aircraft was a modified version of the new production Bf-109 fighter, implying that front-line Nazi military hardware was unbeatable.

This record for piston-engined aeroplanes remained unbroken for 30 years. Finally, on 16 August 1969, great American racing pilot Darryl Greenamyer, flying his award-winning Grumman Bearcat *Conquest I* in four speed runs across a straight 3 km course at Edwards Air Force Base, set a new record at an average speed of 482.462 mph. Subsequently, a few other racers have established even faster speed records.

After the Second World War, air racing took on a very different character. The jet engine changed the air racing equation for ever. No longer would piston-engined aeroplanes vie for the exalted title of world's fastest aircraft.

In 1994, the outskirts of Phoenix became a venue for modern air racing. The annual event, called the Phoenix 500 in deference to the motor-racing world, has been advertised with colourful paint schemes on vintage aeroplanes. The sponsorship of the 1995 races by an emerging chain of appliance stores gets prominent exposure in the event's logo. *(Phoenix, 1995)*

By the war's end, propeller-driven aeroplanes had nearly reached their performance zenith. In fact, the laws of physics intervened, further substantial speed gains being impeded by the newly encountered phenomenon of compressibility. As Second World War fighters such as the Lockheed P-38 Lightning and North American P-51 Mustang approached the speed of sound in dive tests they became barely controllable, at times even breaking apart under stress. Transcending the 'sound barrier' became the object of the new generation of jet aircraft. All that remained for the Second World War piston-engined machines was the chance of attaining modest speed improvements from year to year through a continual series of airframe and engine modifications.

Nevertheless, for a few years after the war the great tradition established at Cleveland was rekindled. In 1946 the National Air Races started again. The war had tremendously advanced aeronautical technology. In the last Thompson Trophy Race before the war, Roscoe Turner won in his Laird-Turner L-RT at a speed of 282.536 mph. Seven years later, when the races resumed, Alvin 'Tex' Johnston came in first flying a Bell P-39Q-10 at a speed of 373.908 mph.

It appeared that the excitement of those halcyon days of the 1930s was being revived. There was a whole new field of racing aeroplanes. Fighters made famous in dogfights over the lush European landscape and the distant Pacific islands came to Cleveland in eye-catching paint schemes. Some of the old-time racing pilots, like Steve Wittman and Earl Ortman, showed up flying these military surplus machines.

In many ways the stage was set for today's air racing. The races seemed to be a contest between the liquid-cooled and air-cooled fighters. On the one hand there were Airacobras, King Cobras, Mustangs, and Lightnings, and on the other there were Corsairs. The top speeds at the races were pushed to new levels, and by 1949 they had almost reached 400 mph.

A notable difference from the pre-war years was that the names of the racing aeroplane manufacturers were now synonymous with vast commercial enterprises. In the old days, racers tended to be the product of prodigious individuals working in borrowed hangar space or a barn next to an improvised airstrip. The racers of the 1930s were distinctive, one-of-a-kind creations.

By the second half of the 1940s the hottest racers were production-line fighters. Some of these were highly modified, but the basic airframes were run-of-the-mill products from the wartime plants of corporations such as Lockheed, North American, Curtiss, Vought, Grumman, Martin, and Douglas. The hallmark of air racing had become refinement of existing speedsters.

In 1948 racing entrepreneur J. D. Reed and talented aircraft designer Walter Beech teamed up to modify a fighter for the Cleveland air races. They decided that an early-model Mustang, a 'razorback' version as opposed to one with a bubble canopy, would be more aerodynamically efficient, so, a P-51C was obtained.

Production Mustangs had a large belly scoop for air cooling. To reduce the drag caused by this protrusion, the belly scoop was eliminated and the cooling apparatus was moved to special wingtip fairings. This was one of the more extreme attempts at aerodynamic streamlining. Called *Beguine* after the catchy tune 'Begin the Beguine', the sleek racer with its bulbous wingtip attachments was painted green, with yellow trim in the form of musical bars and notes. This stunning machine, which seemed to offer so much potential, was purchased by Jacqueline Cochran, the most renowned aviatrix of her time. Although she had campaigned most of her life for gender equality, and had made great strides towards that goal through such accomplishments as helping to organise the women's airforce service pilots during the Second World War, Cochran was precluded from racing her new aircraft because of the then enforced prohibition against female participation in the unlimited closed-course race. Constrained by the ill-conceived rule, Jackie chose her friend, Bill Odom, to fly *Beguine* in the 1949 Thompson Trophy Race.

On the second lap the modified Mustang veered off course and crashed with immense force into a house, killing not only the pilot but a mother and her infant son. A fiery explosion erupted, flames engulfing the house and smoke billowing into the sky.There were unsubstantiated rumors that *Beguine* was modified with one wing shorter than the other to allow it to negotiate the traditional four-pylon rectangular course more easily. According to these reports, the change to a seven-pylon circular course in 1949 made *Beguine* prone to unstable flight. However, the historical consensus is that *Beguine*'s wings were of equal span. The racer probably entered a high-speed stall when Bill Odom recovered from a steep bank around one of the course pylons.

Whatever the case, this tragedy contributed to the end of air racing in Cleveland. There were other factors. Piston-powered,

'Fly Fast, Fly Low, Turn Left' – the simple, accurate, and clever buzzwords of the Phoenix 500 Air Races. At the event, this pithy slogan was imprinted everywhere, enlivening T-shirts, coffee mugs, and, yes, even the press bus. *(Phoenix, 1995)*

propeller-driven air racing in the jet age seemed pointless to many observers, especially given the attendant hazards. Corporate sponsors were pressured to drop their support. The outbreak of war in Korea in June 1950 was the final blow. Some of the racing pilots were called up for duty, and the USA military's customary presence at the air races would be curtailed that year.

During much of the ensuing 14-year air racing hiatus, an energetic and visionary pilot and cattle rancher dreamed of a revival. By 1964, this man, Nevadan Bill Stead, had assembled a wide array of air racing talent at his private airstrip, known as Sky Ranch, located in the sparsely populated area north of Sparks. National air racing was thus reborn in the open spaces of the Nevada desert.

For two years, racers in a variety of classes competed at Bill Stead's Sky Ranch, but the conditions were far from ideal. The runways and parking areas were not paved, and dust was kicked up every time a propeller turned. It became apparent that a new locale was essential, particularly if large audiences were to be accommodated. When word spread that the USAF base ten miles northwest of Reno was about to close, the solution seemed obvious. Coincidentally, this base had been named after Bill Stead's brother, Croston, who crashed at the site while flying a P-51 as an officer in the Nevada Air National Guard.

Sadly, Bill Stead himself perished when his racer crashed during a practice flight in Florida in the spring of 1966. Bill was only 43 years old. Ironically, he never saw the air races take place at the old air base, where they have been held every September since 1966.

Built in the early 1940s, what is now Reno-Stead Airport was used for glider training during the Second World War. At that time, gliders were seen as an expeditious means of transporting large numbers of troops to war zones. Following the war, the base became home to the Air Force's survival school, and also served as a helicopter training site. Indeed, the Nevada Army National Guard is now represented on the field with a contingent of huge CH-47 Chinooks.

The flavour of today's air racing extends from the aeroplanes and pilots to virtually every facet of the sport. Typically, the well financed unlimited teams mark their crew and equipment vans in their well-known racing colours. A huge semi is used by the *Rare Bear* team to transport essential parts, tools, and team members. When not ripping through the sky, the racer is conveniently parked next to the van where pliers and wrenches are within arm's reach. *(Reno, 1995)*

A vestige of Williams' military heritage, a permanent VIP welcome mat sprayed on to the transient ramp now greets Phoenix 500 spectators. Many of America's leading combat aviators and airpower strategists deplaned nearby and strode confidently across this piece of tarmac, an inconspicuous waypoint for aviation history-makers. *(Phoenix, 1995)*

Over the years Reno has grown with some sprawl to the north. Where once there was an almost untouched vista looking out from Reno-Stead Airport, homes and businesses have sprouted up along the corridor marked by the 395 Freeway that winds its way from the city to the airport. There are pockets of development scattered elsewhere throughout the region as the local economy expands in a diversification away from the casino industry. Yet the land in the immediate vincity of the airport is characterised by wide-open expanses dotted with dry lake beds. The surrounding terrain still makes it a good air racing site. The field elevation, at 5,046 ft above sea level, is also a plus because the unlimited racers perform better at such an altitude.

Today, pylon-style air racing occurs in half-a-dozen different classes: the unlimiteds, comprising the fastest aeroplanes, generally Second World War fighters (the organisers of the Phoenix air races have split this class into two subsets – the professional racers, which are significantly modified for racing, and the vintage stock racers, which remain virtually unchanged from their original configurations); the North American AT-6/SNJ trainers; International Formula One aircraft, which are small aeroplanes designed and built expressly for racing and are required to use a 200 in^3 displacement engine; sport biplanes; T-28 trainers; and helicopters. The latter two classes are recent creations of the organisers of the Phoenix air races.

The biggest race-course is that for the unlimiteds. It measures 9.128 miles in an imperfect oval shape. The courses for the other classes of racers are overlaid on the grounds of the unlimited course. The AT-6s fly a near-rectangular course 4.99 miles long. The Formula One and Biplane Classes fly the same course of 3.1068 miles. Each course is strategically marked by pylons. Depending on the class and the heat or race, the number of laps flown ranges from five to eight.

No racer can dip below a pylon nor climb above 1,500 ft except in an emergency. Cutting inside pylons is prohibited as well. The Unlimited and AT-6 Class races are started in the air, whereas the Formula One and Biplane Class races have a standing 'racehorse' start. The contenders must complete the requisite number of laps for their given race. All finishers cross the home pylon and are then given a wave of the chequered flag.

Qualification trials are held in the days leading up to the races. A series of heats are then conducted in each class to determine the line-ups for the various races. Each class has Bronze, Silver, and Gold races. Similarly matched aeroplanes fly in these races, with the fastest competing in the Gold race.

In 1994 full-blown annual air races were initiated outside Phoenix at the Williams Gateway Airport, previously Williams Air Force Base. So that this event did not conflict with the established Reno event held each September, the Phoenix air races were planned to occur in the spring of each year. The organisers of the Phoenix air races reasoned, quite logically, that because the event took place on the periphery of a major urban centre the attendance was likely to exceed the turnout of about 50,000 spectators on the prime days at the Reno air races. On the other hand, the burgeoning development in the Phoenix area presents some drawbacks.

The crash of the *Super Corsair* within the boundaries of the adjacent General Motors proving ground during the 1994 Phoenix air races raised some eyebrows. Although no one on the ground was injured, in 1995 the course for the Unlimited Class Races was clipped to 7½ miles to reduce the chances of that possibility occurring. With a shorter unlimited course, along with a field elevation of 1,383 ft above sea level, this has made some observers wonder whether the fastest unlimited racers at Phoenix will be precluded from reaching the highest speeds attained at the Reno air races.

When Williams was an Air Force Base, Service personnel affectionately referred to it as 'Willie'. Built as a pilot training establishment early in the Second World War, Willie continued its educational mission until it closed as a military installation in the autumn of 1993. Its infrastructure includes three long parallel runways, massive concrete ramps, and numerous buildings scattered on the expansive grounds. Like Reno, the ceiling and visibility at Phoenix are usually ideal. Temperatures are known to exceed the comfort level in the middle of the day, and the intense heat bouncing off the Arizona desert floor can produce choppy air for the racers as they skim low over the surface in their contests.

Promoted under the Phoenix 500 Air Races banner, the event's organisers have borrowed from the world of automobile racing in an attempt to duplicate the success of motorsports. Innovations at the Phoenix air races include the use of giant electronic screens, like those that baseball fans are accustomed to seeing, advertising a variety of products between postings of race line-ups and results. Not missing an opportunity, the organisers offer an evening musical concert as part of the entertainment package.

While newcomers to the world of air racing promotion and an expansion of air racing venues are welcome, air racing's milieu for more than the last three decades has been distinguished by the hilly buckskin landscape northwest of Reno, Nevada. Unpretentious but proud, the world's racing pilots bring their amazing aeroplanes to this dusty enclave every September in a continuation of the glorious tradition begun in Cleveland long ago.

The calm of the desert sunrise, its indigo sky gradually softening with the rugged topography cast in purplish hues, awakes to the stir of the wind. The pungent fragrance of the flowering sagebrush sweeps across the flat valley, permeating the lucid air. Neatly delineated trails of dust arise, signalling the day's first arrival of humans. They are the pylon judges, the umpires of air racing, on the way to their remote stations on the airport grounds. In a little while the freeway traffic from town begins to build. A purposeful quiet descends upon the pit, air-racing crews addressing last-minute maintenance concerns, tightening bolts or pumping up tyres.

About the time that the grandstands reach the saturation point, the thunderous herd of unlimited racers enters the beckoning sky. Minutes later, the first hint that these airborne monsters have joined in formation comes from the crackling chatter emanating from the portable radios on the airfield. Veteran air-racing observers, knowing that the gaggle of aeroplanes is flying relatively low and that the radio transmissions are limited to line of sight, gaze to the south, in the direction of Dry Lake Summit and Peavine Peak.

Then, in the distance, a row of transient specks comes into view. A lone dot, the pace aeroplane, maintains its position above and behind the rest. From the radios' speakers the voice of the pilot in the pace aeroplane gently cajoles: 'Number Five, pull it up.' 'Number Twenty-eight, hold steady.' 'Everybody come in a little tighter.'

Still too distant to identify by make and model, but close enough to recognise as vintage aeroplanes, the racers are screaming down the chute. They must remain evenly aligned at this critical juncture, or the start will be aborted. Each pilot has one hand on the throttle, itching to plunge it forward; and the other on the control column, keeping the aircraft in position. They are intensely focused on the other racers, ensuring proper separation.

The pace pilot's voice chimes in again: 'Gentlemen, you're lookin' good. Hold it right there'. On the ground, apart from the radio communications, there is hardly a whisper as everyone, rapt with anticipation, strains to see the broadening armada of mighty racers.

One more reassurance over the radio as the start of the race nears: 'Gentlemen, you're lookin' good'. Closer and closer they come.

Then, in a departure from the casual, matter-of-fact intonations expected of pilots, and as if knowing he is entitled to a great flyer's poetic licence, courageous military aviator, brilliant test pilot, and extraordinary air show performer Robert R. A. 'Bob' Hoover shouts out in a breathless laconic flurry: 'Gentlemen, you have a race!'

The pace aeroplane zooms virtually straight up, marking its vertical path with a plume of white smoke. The racers, now unleashed, break from the starting formation and streak furiously for the pylons in a loud clamour that has become a ritual of this otherwise serene place in the desert. The custom lives on; the air races have started!

At the former Williams Air Force Base on the southeast fringe of the Phoenix metroplex, the participants and spectators at the Phoenix 500 are never far from reminders of the airfield's past. Established in the early days of the Second World War, 'Willie', as the base was fondly known, became the largest of the Air Force's undergraduate pilot training centres. Williams Air Force Base is where tens of thousands of student pilots earned coveted silver wings on Cessna T-37 'Tweet' and the Northrop T-38 Talon trainers. It is now called Williams Gateway Airport, and the local community is trying to promote commercial activities at the enormous site, which ceased operating as a military establishment in 1993 owing to Pentagon budget cuts. Pedestal-mounted training jets stand guard near the old base headquarters building. *(Phoenix, 1995)*

Appropriately, the 152nd Reconnaissance Group of the Nevada Air National Guard, known as the 'High Rollers', perform a formation fly-over to mark the official start of the Reno air races. In 1948 the Nevada Air National Guard was established at Reno Air Force Base, now called Reno-Stead Airport and the site of the racing action. Eventually the Guard moved to the larger commercial airport in Reno, but it maintains a recurring presence at the annual races that are held at its birthplace. For 20 years the 'High Rollers' flew the McDonnell Douglas RF-4C Phantom II, but aircraft age and budget considerations forced the type's retirement shortly after the 1995 Reno air races. The unit was scheduled to convert to Lockheed Martin C-130 Hercules transports for statewide firefighting missions. *(Reno, 1994)*

Chapter 2
Show Time

Air racing and air shows go hand-in-hand, for racing pilots, though they may deny it, are performers just like air show pilots. Indeed, some racing pilots, such as Marvin 'Lefty' Gardner, are admired as much for their gutsy pylon hugging as for their graceful exhibition flying. From the early days at Cleveland, air racing fans were entertained with wingwalkers, formation teams, and aerobatic routines. This tradition continues at today's major air races.

The air racing audience is treated to spectacular air show performances between the races. The 'Red Baron' Stearman Squadron, composed of four Stearman biplanes of Second World War vintage, executes a precisely symmetrical formation loop. Each aircraft has been modified with a 450 hp Pratt & Whitney engine and a constant-speed propeller, and when the team passes overhead the sound is music to the enthusiast's ears. Veteran team member John Bowman piloted the lead aircraft during the 1995 season. *(Reno, 1995)*

Once in a while, during a lull in the races, an unannounced fly-by occurs. Here a Rockwell B-1B Lancer, its variable-geometry wings swept fully aft for fast cruise and its bomb bay doors open to simulate a bomb run, sweeps past the awed spectators. *(Reno, 1995)*

Delmar Benjamin's beautifully constructed replica of the Gee Bee R-2 takes the audience to that fabled era of the 1930s, when daring pilot Jimmy Doolittle, in one of the original Gee Bee models, set a new speed record at the Thompson Trophy Race in Cleveland. A masterful pilot, Delmar pushes the fat, bumble bee-shaped aeroplane to the edge of its envelope, rolling knife-edge and then inverted for prolonged passes with the Gee Bee's tail seemingly only inches from the ground. Delmar and the Gee Bee are guaranteed crowd pleasers. *(Reno, 1995 & Reno, 1994)*

The Canadian Forces 'Snowbirds', officially known as 431 Air Demonstration Squadron, celebrated their 25th anniversary year in 1995 by giving a few select performances south of the border, including the Reno air races. Flying Canadair CT-114 Tutor trainers, the nine-aircraft team forms a trademark diamond shape. Two jets then break off for solo routines, leaving their seven teammates to execute graceful formation manoeuvres. In their half-hour display the 'Snowbirds' perform about 50 manoeuvres. Under the command of Major Steve Hill for the 1995 season, the team's precision and showmanship were unsurpassed.
(Reno, 1995)

A contender for the Biplane Gold in 1995 was *Patty Anne*, a Pitts Special flown by Yucaipa, California, resident Michael Stubbs. *(Reno, 1994)*

OPPOSITE:
Although their racing speeds do not reach half that of the fastest unlimited racers, the biplanes are nevertheless usually locked in tight heats, with the pilots clinging to the pylons and jockeying for position so as to eliminate any unneccesary time around the circuit. *(Reno, 1995)*

Not all racing aeroplanes are speed demons. The Biplane Class consists mainly of various Pitts models. Known more for their aerobatic prowess, these nimble little biplanes, some incorporating streamlining modifications, have approached 200 mph in the fiercely competitive races. *(Reno, 1994)*

Chapter 3
Two Wings at a Time

Biplane racers virtually disappeared after some stunning victories in the Thompson and Bendix Trophy Races in the early and mid 1930s. Aeronautics was advancing, and metal airframes, cantilevered wings, and enclosed cockpits were taking over as the racing biplane slid into history. Yet there has always been an alluring mystique surrounding aircraft with two wings. The glory of the old-time racers continues in the modern Sport Biplane Class.

A dogfight between Pitts biplanes. In the lead is *Ole Yeller*, flown by retired airline pilot Ray Kraspovich of Arvada, Colorado, with *My Pitts*, piloted by South Africa Airlines captain Levin Scully, closing fast. Captain Scully, a first-time participant at Reno in 1995, did surprisingly well.

The winner of the Biplane Class Race at Reno in 1995, with a speed of 202.124 mph, was Patti Johnson, an aerobatic flight instructor from Edgewater, Florida. In 1993, piloting her Mong, named *Full Tilt Boogie*, she became the first woman to win a racing-class championship. In so doing she has admirably built upon the traditions of predecessors such as Laura Ingalls, Louise Thaden, Elinor Smith, Pancho Barnes, Jacqueline Cochran, and Amelia Earhart. *(Reno, 1995)*

At times the naming of the racing aircraft is as creative as the flying techniques used to whittle a few seconds off elapsed time. Called *Moving Violation*, this Pitts Special was flown by B. L. 'Mick' Richardson, a dentist from Auburn, California. *(Reno, 1994)*

Chapter 4
Small but Fast

Limited to a certain type and size of engine, the Continental O-200, the International Formula One Class racers are relatively free of other regulatory requirements. They are designed and built strictly for speed. Accordingly, these tiny, some would say midget, racers are in many ways a springboard to improving aerodynamic efficiency in other classes of racers and in aeroplanes generally. This class boasts the oddest shapes and configurations. Also, for weight control and strength, composite materials are used extensively. The Formula One Class is the experimenter's bailiwick.

As a rule, Formula One aircraft are faster than the biplanes. Obviously there is less drag owing to the absence of a second wing, but there also tends to be a great deal of design innovation. It is common to see some peculiar shapes in this class. This is California architect Kevin Kelly's *Barbara Jean* at Reno in 1994. The following year the tail had been redesigned so that the fin and rudder were below the fuselage. After modification the aeroplane's name was changed to *Barbara Jean 2*. *(Reno, 1994)*

For convenience, most Formula One racers are towed to and from the runway. Safety dictates that they must not be taxied through the crowd line. This is Nampa, Idaho, resident Patrick Rediker's colourful red and yellow Shoestring, known as *Spud Runner*. The equally colourful turquoise pick-up truck serving as tug is of uncertain vintage. *(Reno, 1994)*

Ray Cote's well-known OR-71 *Alley Cat* finished second at Phoenix in 1995 with an impressive speed of 236.535 mph. Despite the competitive spirit on the racecourse, a genuine camaraderie pervades the confined quarters of the race hangar, which is shared by both the Formula One and Biplane Class racers. There is no hesitation among opposing teams in lending a helping hand. Co-operation with a jeans-and-T-shirt casualness is the order of the day for these crews. *(Phoenix, 1995)*

Texan Bruce Bohannon flies the very hot Miller Special M-105 *Pushy Galore*. Generally acknowledged as one of the most formidable racers in its class, this innovative pusher design usually finishes second or third in the Gold races at Phoenix and Reno. *(Phoenix, 1995)*

George Budde of Midwest City, Oklahoma, calls his Cassutt IIIM *Oakie Streaker.* As the petite racer is pushed into the early morning sun, the vast ramp at the old Williams Air Force Base envelops it in a sea of concrete slabs. This aeroplane came in third in the Silver race at Phoenix in 1995. *(Phoenix, 1995)*

With its canopy cover removed, *Oakie Streaker* is towed back to the hangar after a punishing few minutes spent feverishly rounding pylons. *(Reno, 1994)*

This Cassutt, clearly a very lightweight racer, gets a lift as it proceeds by means of a very primitive form of transportation from the hangar to the flight line.
(Reno, 1995)

Retired airline pilot Raymond Sherwood's Cassutt *Silver BB* makes its leisurely way to the races. *(Reno, 1995)*

Ray Cote's *Alley Cat*, racer No. 4, took fourth place in the Gold race. Despite the stress as pilots strive for position, racing can be loads of fun, and you can see a smiling face in the cockpit under the visor's glare. This El Cajon, California, based OR-71 achieved a race speed of 225.147 mph at Reno in 1995. *(Reno, 1995)*

The all-red *Scarlet Screamer* Cassutt of Scotty Crandlemire turns a pylon. Note that the pilot's eyes are focused to the left of the direction of flight, trying to squeeze in as close as possible to the pylon without cutting inside and suffering a points penalty. *(Reno, 1995)*

GERONIMO
96

OPPOSITE:
Katharine Gray, a flight instructor from Fresno, California, one of the growing number of female pilots on the race circuit, flew her GR-7 *Geronimo* to a finish just behind *Alley Cat* at a speed of 223.926 mph at Reno in 1995. *(Reno, 1995)*

A common aircraft type in the Formula One Class is the Cassutt. Raymond Sherwood flies *Silver BB* and fellow Californian Rick Todd pilots *Knotty Boy*. *(Reno, 1995)*

Ray Cote in *Alley Cat* about to overtake Kevin Kelly in *Barbara Jean 2*. Ray won the Silver at Reno in 1995. Kevin claims that his design's odd tail configuration provides the best rudder position in relation to the aeroplane's low wing when turning. Most Formula One racers are of mid-wing design with conventional tails. In 1995, another Kelly-designed F1D entered competition. Called *Smiddy's Firefly* and flown by Steve Myers, it won the Silver at Phoenix and came in sixth in the Gold at Reno. *(Reno, 1995)*

The two pushers, locked in a showdown lap, with bright yellow *Pushy Galore* holding the lead against James Miller in the Miller Special JM-2 *Pushy Cat*. Although fast in its own right, *Pushy Galore* was a distant second to *Nemesis* at Reno in 1995 with a speed of 232.397 mph. *(Reno, 1995)*

The longtime Formula One champion is Jon Sharp, an aerospace engineer from Lancaster, California, a hotbed of aeronautical technology and flight test. His one-of-a-kind Sharp DR-90 *Nemesis* set a new race record for the class at Reno in 1995 with a speed of 249.904 mph. Each season, Jon and his talented team seem to improve this magnificent design in some small way which bolsters their race results. Before the 1995 racing season, the engine air inlets and the oil cooling inlets and outlets were further streamlined, which enhanced performance. With such ongoing innovation there is no telling how long *Nemesis* will dominate the class.
(Reno, 1995)

Racer No. 7, *Yankee Air Pirate*, was flown at Reno in 1995 by Tom Dwelle, Jr., known as 'T. J.'. Competing against him was his father, Tom Dwelle, Sr., flying racer No. 30, *Archimedes*. Father and son displayed no sentiment during their heats, T. J. edging Tom out in the AT-6 Bronze race. T. J. came in second with a speed of 203.697 mph and Tom, who finished third, followed by a whisker with a speed of 203.600 mph. Earlier in the year Tom had lost four fingers in a dreadful accident while working on his unlimited racer, *Critical Mass*. Showing incredible perseverance, he was flying again only three weeks after sustaining the injury, and his indomitable spirit, reflecting his tough-as-nails fighter pilot background, was apparent on the course. *(Reno, 1995 & Reno, 1994)*

Chapter 5
Big and Noisy

The AT-6 Class is confined to the hefty and loud advanced trainers of Second World War vintage. These aircraft are so evenly matched that the races frequently have heartstopping finishes. Unlike the smaller racing aeroplanes, whose races are started motor-racing style on a runway with the waving of a flag, the AT-6s and SNJs have an air start. Once they are aligned near the course, a pace aeroplane calls the start of the race.

The 'pilot maker' was never intended to be a racer. Built in time for the Second World War, the Texan (known throughout the British Commonwealth as the Harvard) was one of the Allies' most prolific advanced trainers. Designated AT 6 by the Army and SNJ by the Navy, this sturdy North American Aviation product gave cadets necessary flying skills after they had learnt the fundamentals of airmanship in relatively docile primary and basic trainers such as the Boeing Stearman and Vultee Valiant. In the Texan, cadets confronted more power in the form of the 600 hp Pratt & Whitney R-1340 engine and complex systems such as flaps, retractable undercarriage, and navigational/communication radios. It was said that if a student could fly this aeroplane he could fly anything. The AT-6 was truly a pilot maker. *(Reno, 1995)*

Enjoying a prize parking spot next to the start-up ramp is the 1992 AT-6 Class winner, *Warlock*. Belonging to veteran racing pilot Alfred Goss, a cropduster from Woodland, California, this beauty has its racing history, with dates and speeds, painted on the fuselage under the cockpit. *(Reno, 1994)*

For every eight minutes or so of pylon racing at Reno, the aged trainers require many hours of arduous maintenance. Here, the cowling is removed to allow access to the engine of *Boomer*, the proud racer of Joe Hartung. In 1994 Joe came in third in the Gold race. *(Reno, 1995)*

One of the sharpest-looking AT-6s at Reno in 1995 was airline pilot Bud Granley's *Four Play*. Even among the creatively decorated aircraft at the races, this machine, with its combination of polished metal and red/yellow paint scheme stood out as a showstopper. *(Reno, 1995)*

OPPOSITE TOP:
Happily soaking up the desert sun was Keith McMann of Delta, British Columbia, and his Harvard Mark IV (a Canadian version of the AT-6) *Red Knight*. A knight's helmet is emblazoned on the cowling.
(Reno, 1994)

OPPOSITE BOTTOM:
For years, famed racing pilot Eddie Van Fossen, now retired, and his *Miss TNT* dominated the AT-6 Class. Indeed, they captured seven national championships – an all-time record. In 1994, challenger Charles Hutchins finished second in the Gold race at a speed of 221.636 mph. Here, his crew can be seen towing his all-blue *Mystical Power* to the start-up ramp.
(Reno, 1994)

Texas dentist Jim Bennett flies the well-known *Tinker Toy*, which finished second in the Silver race at Reno in 1995.
(Reno, 1995)

ROYAL CANADIAN AIR FORCE
64
20264

CHARLES HUTCHINS
JIM WOODFORD
21
N125JD

In 1994, tragedy befell the National Championship Air Races at Reno. At the outset of the AT-6 Silver race on 18 September, just as the pace aeroplane called the start and pulled up, two aircraft collided. Before a horrified crowd one of the racers began to break up in the air, its parts tumbling to the desert floor.

Apparently, the tail of Ralph J. Twombly's racer No. 41, *Mis Behavin'* scraped against the outer portion of the lower left wing of Jerry McDonald's racer No. 5, *Big Red.* Pieces of No. 41's tail broke off, rendering the aeroplane uncontrollable. It pitched up and rolled over No. 5, the stress presumably causing the left wing to separate. The bulk of No. 41 that remained in one piece, the fuselage and right wing, rotated awkwardly for a few moments before hitting the ground. Various pieces of debris, including the remnants of the left wing, continued to float to the ground for some seconds, appearing like so much *papier mâché* to spectators in the distance. No one on the ground was hurt.

In an amazing feat of airmanship, Jerry McDonald managed to nurse his damaged No. 5 on to one of Reno-Stead's runways without further incident. The aeroplane's left aileron was gone, and there was damage to the left wingtip and part of the right wing. Jerry was determined to carry on after his frightening experience. *Big Red* was repaired, and he came back to win the Silver race at Reno in 1995.

Sadly, one of air racing's greatest pilots, Ralph Twombly, perished in the accident. A commercial pilot with more than 30,000 flight hours to his credit, Ralph had become a part of the Reno scene, participating in the air races regularly from 1969. He had won the AT-6 championship in 1977 and 1982, and had even competed in the Unlimited Class in 1989 and 1990. Ralph, from Wellsville, New York, was 67 years old. Those who saw Ralph competing over the years at Reno could tell that the sky was his element. A true aviator, he left an outstanding legacy and is sorely missed. *(Reno, 1994)*

The AT-6 Gold champion in 1994, Eddie Van Fossen, an experienced cropdusting pilot from Bakersfield, California, approaches a pylon in his *Miss TNT*. The winning speed was 224.704 mph.
(Reno, 1994)

Third place in the 1994 Gold race was taken by Joe Hartung, flying *Boomer* at 220.462 mph. He can be seen here peering over the instrument panel, leaning towards the pylon. *(Reno, 1994)*

A perennial favorite of air racing fans, Jim Bennett's *Tinker Toy* zooms around the pylons with Tom Dwelle, Jr. at the controls. The alternating black and white stripes emanating from near the wing roots are a variation on D-Day invasion markings. *(Reno, 1994)*

Called *Slo Yeller*, this Texan is piloted by Dorel Graves, Jr. *(Reno, 1995)*

Sporting a new paint scheme for 1995, Gene McNeely's *Undecided* had a decidedly fast finish as number three in the Gold race at a speed of 221.441 mph. *(Reno, 1995)*

Clouds occasionally develop over Reno-Stead Airport, but as long as the ceiling does not fall below minimums the races continue. This Texan, *Killer Unit*, is flown by California airline captain Ralph Rina. *(Reno, 1995)*

Banking steeply as it rounds the pylon, Al Goss's *Warlock* keeps its competitive edge. At a speed of 221.046 mph, racer No. 75 managed to secure fourth place in the 1995 Gold race. *(Reno, 1995)*

Little wonder that John Luther of Bulverde, Texas named his racer *Texas Red*. Indeed, it was hot in the 1995 Gold race, coming in sixth at a speed of 216.991 mph. This is the ultimate Texan, blending marvellously with the crisp blue Reno sky. *(Reno, 1995)*

The view from the pylon as *Warlock* rushes ahead, throttle nearly at the stop, all nine Pratt & Whitney pistons hammering away, and propeller blade-tips nearing supersonic speed. The desert comes alive with the thumping clatter of low-flying racers. *(Reno, 1995)*

The 1995 AT-6 Class Gold Race champion was Charles Hutchins of Seabrook, Texas, in *Mystical Power*. With a speed of 231.430 mph, Charles crossed the finish line a mere 0.21 sec (about two-tenths of a second) ahead of his nearest challenger. The AT-6 Class often provides really breathtaking races, as the aeroplanes are so equally equipped. The clean lines of the AT-6/SNJ can be seen from this viewpoint. Although it was not designed for racing, the type makes air racing exciting. *(Reno, 1995)*

Big Red and *Tinker Toy* jockey for position low over the Nevada desert. *(Reno, 1995)*

OPPOSITE PAGE:
A different view of the Texan shows the exposed wheels in the retracted position. Many AT-6/SNJ racing pilots like to decorate their aircraft in historic colour schemes. This example, named *Archimedes*, is finished in the colours of an SNJ based at the old Naval Air Station at Grosse Ile, Michigan, an island outpost where the Detroit River joins Lake Erie. Today, the airfield is a general aviation airport. *(Reno, 1995)*

Even the pace aeroplanes can hardly resist rounding some of the pylons after the races. This Texan, from the collection of the Cavanaugh Flight Museum of Dallas, makes a 'clean-up' circuit of the course, ensuring there are no stragglers. *(Reno, 1995)*

NAVY
302
GROSSE ILE
NAVY
GROSSE ILE

With Nevada's ubiquitous buckskin mountains as a backdrop, a gaggle of AT-6/SNJ racers wind their way around the pylons as they enter the course from the chute. In the lead is *Mystical Power*, the 1995 Gold race champion, followed by *Undecided*, *Four Play*, and *Wyoming Wildcatter*. *(Reno, 1995)*

Father and son race against each other. Tom Dwelle, Sr., in *Archimedes* was edged out by Tom 'T. J.' Dwelle, Jr., seen in the lead here, flying *Yankee Air Pirate*. T. J. came in second in the 1995 Bronze race at a speed of 203.697 mph, followed by his father, who was third at a speed of 203.600 mph. That is close! *(Reno, 1995)*

With the aircraft barely a heartbeat from each other, this is no time to sneeze or hiccup. Adrenalin is really flowing now, as Gene McNeely tries to maintain his lead over Bud Granley and James Good. *(Reno, 1995)*

This looks almost like formation practice. In traditional markings and in another era it might well have been, but at the moment these aeroplanes are being pushed for every ounce of energy. *Wyoming Wildcatter* bettered *Four Play* and went on to take fifth place in the 1995 Gold race at a speed of 218.593 mph. *(Reno, 1995)*

Chapter 6
Whirling Racers

Helicopter pylon racing was a novel idea which came to fruition at the 1995 Phoenix 500. Tipped on their sides, with rotor blades spinning, these eggshell-shaped competitors appeared as windmills. As with their fixed-wing brethren, speed was the prime consideration.

For the first time at the Phoenix 500 in 1995, pylon-style helicopter races occurred. The impetus for this new form of racing was the friendly rivalry of two local helicopter pilots. This developed into an invitational race in which five experienced operators competed in McDonnell Douglas MD-500 choppers. Flying racer No. 3, the MD-500D *Bad Attitude*, was Thomas Hauptman, also a Formula One pilot. Here, the competing helicopters are prepared for a heat. *(Phoenix, 1995)*

Thomas Hauptman, a highly regarded helicopter operator in Hawaii, blasts around the pylons. *(Phoenix, 1995)*

Scott Urschel, the owner of Scottsdale, Arizona, based Urschel Manufacturing Industries, brought his MD-500C to the races for local helicopter flight instructor Neil Jones to fly. It has been re-engined with a newer, more powerful 420 shp engine common to the D and E model MD-500s. This earlier version of the MD-500 has a four-bladed rotor, while the newer models have five-bladed rotors.
(Phoenix, 1995)

Once they have been towed a safe distance from the parking ramp and the spectators, the helicopters fire up and assemble elsewhere over the airport property for the hover start of their race. Pole position was assigned on the basis of performance in the preliminary heats, much as it is done with fixed-wing racers. In the foreground is Mike Dillon's helicopter, while first off the ground is Scott Urschel in a very modern MD-520N 'Notar' leased from McDonnell Douglas Helicopter Systems, based only ten miles away at Falcon Field in Mesa. The Model 520N has no tail rotor, hence its name. A compressed-air system in the tailboom provides the anti-torque function. Mike and Scott were the two helicopter pilots whose friendly rivalry inspired the inauguration of these races. *(Phoenix, 1995)*

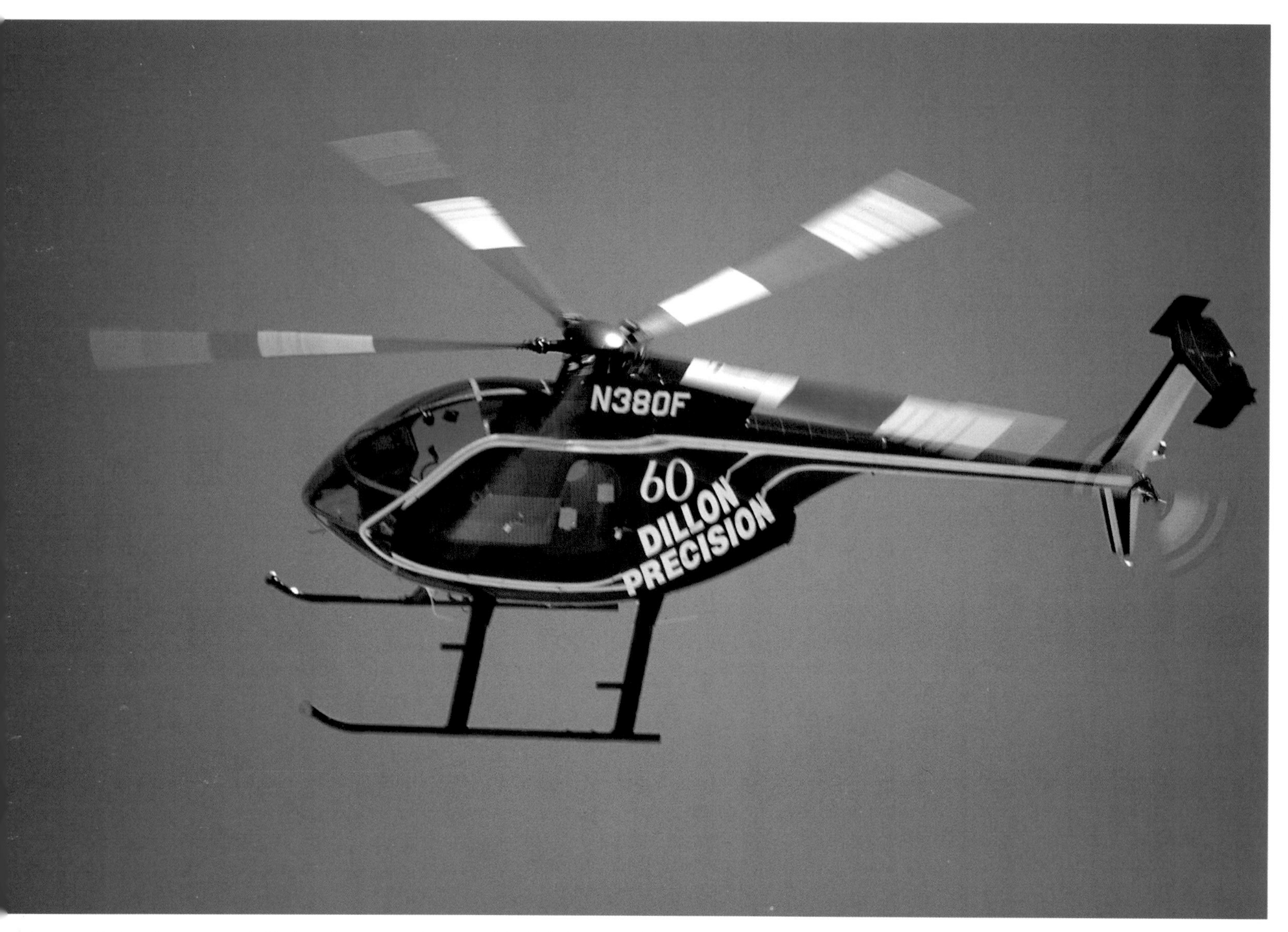

Returning to level flight after rounding a pylon, Mike Dillon adds power, adjusting cyclic and collective, as he tries to take advantage of the straightaway.
(Phoenix, 1995)

Faint hearts skipped at least a few beats as the rotor blades of competing helicopters appeared to get within close proximity of each other in a clearly-fought contest reminiscent of the famous chariot race in the Hollywood epic *Ben Hur*. Thomas Hauptman and Mike Dillon ran neck-and-neck at this point in the race. Tom captured the Gold with a speed of about 145 mph; pretty fast for a helicopter flying laps on a closed course. *(Phoenix, 1995)*

Chapter 7
Big and Noisy II

The North American T-28 Trojan, a US military advanced trainer of 1950's vintage, had appeared in races with the unlimiteds over the years, but it never really measured up to the much more heavily powered fighters. The organisers of the Phoenix 500 recognised this, and, faced with an absence of AT-6s at their event, decided to establish a new racing class just for T-28s.

In place of the AT-6 Class, a whole new race category, the T-28 Class, was created for the 1995 Phoenix 500. Featuring only the Navy's B model for competitive reasons (the A was a lower powered Air Force model and the C was a weighted down Navy model for carrier qualifications), the T-28 Class promised to be an exciting addition to the air racing circuit. *(Phoenix, 1995)*

The T-28 belonging to Neil Anderson, former Marine Corps fighter pilot and chief test pilot for Lockheed Martin's Fort Worth Company, stands on the ramp awaiting race time. *(Phoenix, 1995)*

T-28s were among the last radial-engined aircraft in the US military's inventory. Designed as trainers, some variants were used for ground-attack missions in Vietnam. This is the T-28 flown by Boyce Thelen of Loomis, California. *(Phoenix, 1995)*

Bruce Wallace of San Carlos, California, brought this aircraft to the first T-28 Class race. *(Phoenix, 1995)*

NAVY
095
NJ
95
VA122
NAVY
138347

USAF

Bruce Wallis banks his T-28B as he approaches one of the pylons. *(Phoenix, 1995)*

OPPOSITE PAGE TOP:
Sometimes referred to as flying washing machines, T-28s are known for producing a raspy sound. These big, heavy aeroplanes with their mighty 1,425 hp Wright R-1820 engines make quite a racket in the sky. *(Phoenix, 1995)*

OPPOSITE PAGE BOTTOM:
Developed between the Second World War era Texan and the turbine-powered trainers of today, the T-28 has a performance akin to that of its older brother, the AT-6. The winning speed on the course at Phoenix in 1995 was about 239 mph, a mere 8 mph faster than the 1995 Reno AT-6 Class champion. *(Phoenix, 1995)*

Evenly matched, like the AT-6 Class racers, these T-28Bs are piling the coals on the fire as they scramble for position. The winner of the Gold race was San Diego, California, resident Jay Shower in *The Bear*. *(Phoenix, 1995)*

An unusual sight in the Nevada desert; a former Royal Air Force jet trainer/light-attack fighter is parked next to the Unlimited racers. This Folland Gnat Mk 1 served for the first time as the Unlimited Class pace aeroplane and safety observer at the 1995 Reno air races. Piloted alternately by Skip Holm and Dean Cutshell, the backseat passenger was the legendary Robert R. A. 'Bob' Hoover. For many years, Bob flew his P-51 *Ole Yeller* as the pace aeroplane and signalled the start of the races with his famous line: 'Gentlemen, you have a race!' Having been unable to perform that role for a few seasons because of the loss of his US medical certificate, Bob was given the honour of calling out the start of the races from this special perch. *(Reno, 1995)*

Chapter 8
The Roar of Thunder

Ripping low across the desert, the Unlimiteds, almost all Second World War era fighters, sweep around the pylons as fast as piston powered, propeller-driven aircraft possibly can. Many of the old warbirds have been modified with lowered cockpit canopies, clipped wings, high-tech engine cooling, and improved powerplants. From time to time, scratch-built racers like *Tsunami* and the *Pond Racer* have been entered in the Unlimited competition, offering expectations of new records. Unfortunately, these have been lost in accidents after short racing careers.

The quest for higher speeds is insatiable. Over the years, record speeds for the Unlimited racers have edged up, sometimes in dramatic leaps and at other times in small increments. From the end of the Thompson Trophy Races in 1949, when the racing aircraft were the same basic aeroplanes as used now, to the present day, the recorded racing speed for the Unlimited champion has improved by about 70 mph. This represents an average annual increase of 1½ mph over the last 46 years. While the rate of increase has clearly tapered off since the immediate post-war years, the push for record speeds continues.

The owner of this magnificent Lockheed P-38L Lightning, named *White Lightnin'*, is hardly superstitious. The use of the number '13' is, in fact, a rebuff of unfounded myth. Beloved racing pilot Marvin 'Lefty' Gardner has not only dazzled the crowd with his pylon hugging in competition, but has impressed aviation aficionados with the smoothness of his aerobatic routine during the air show portion of the Reno event. *(Reno, 1994)*

While the air race spectators are back at the casinos in downtown Reno, the crews are busy tending to their precious aircraft. The aeroplanes accumulate dirt, dust, and lubricant residue in the few minutes spent straddling the pylons, but the devoted technicians who keep the finely tuned machines humming, dare not spill a drop of fluid during maintenance. *(Reno, 1994)*

One of the more colourful participants at Reno over the years has been this candy-apple red, early model Mustang named *The Believer.* *(Reno, 1994)*

The highly modified, gargantuan Hawker Sea Fury once known as *Blind Man's Bluff* was resurrected as *Critical Mass* by Tom Dwelle, Sr. It strikes an intimidating stance in the Nevada high desert. *(Reno, 1994)*

A full-sized moving van is necessary to support an Unlimited Class racer like this North American P-51 Mustang *Huntress III* flown by airline pilot Robert Converse of Santa Paula, California. *(Reno, 1994)*

Buttoned up again, *Section Eight* looks like a sparklingly restored relic of the Second World War, replete with victory tallies and D-Day invasion stripes. This is actually a versatile paint scheme, for it allows the aeroplane to receive preferred parking at warbird air shows without impairing its ability to be employed as an Unlimited racer. *(Reno, 1994)*

Considered a pre-eminent Allied fighter of the Second World War, the P-51 embodied much to endear it to the pilots who steered it through combat. Manoeuvrable and long-legged, reliable and fast, durable and lethal, this aeroplane with its majestic presence has imprinted itself on the imaginations of aviation enthusiasts everywhere. One can hardly walk the air racing ramp without turning to gaze at these remarkable specimens. *(Reno, 1994)*

The fabled *Dago Red*, a highly modified P-51 and past Unlimited champion, receives tender care under the tutelage of members of the Santa Monica, California, Museum of Flying. *(Reno, 1994)*

The highly publicised contest between *Strega* and *Rare Bear* highlights an argument as to which aircraft type makes the better racer: the sleek aeroplane with a liquid-cooled engine or the fat aeroplane with an air-cooled engine. Since the results tend to seesaw back and forth over the years, there is really no definitive answer. As *Strega* prepares for a heat, no task is too small for the maintenance crew when making the final checks before flight. *(Reno, 1994)*

One of the official pace aeroplanes for the Phoenix 500, this P-51, labeled *Spirit of Phoenix*, was leased by the Planes of Fame Museum of Chino, California, to the Superstition Racing Corporation. It was piloted in 1995 by Kevin Eldridge, who baled out of the disabled *Super Corsair* at the previous year's Phoenix air races. (Phoenix, 1995)

Among the more effervescent paint schemes to come on the scene in 1995 was this job on Delbert Williams' *Voodoo Chile.* (Phoenix, 1995)

GMC
CALIFORNIA
4Y58021
76
31

"MILLION DOLLAR BABY"
SACRAMENTO JAGUA
MILLION DOLLAR BABY"
MUSTANG

Perhaps the most unusual racer in the Unlimited Class is Floridian Don Whittington's P-51, fitted with a Rolls-Royce Griffon engine driving contra-rotating propellers. The theory is that this arrangement neutralises torque, channeling more of the power into useful energy. The larger Griffon replaced the Mustang's normal Merlin engine. Obviously, the aircraft itself has undergone aerodynamic modifications. This configuration has had some teething problems, but offers great potential. *(Phoenix, 1995)*

PREVIOUS PAGES:
Art Vance, Jr., an airline pilot from the small California town of Sebastopol, flies *Million Dollar Baby*. The external fuel tank in the foreground has had wheels and a couple of open cockpits installed for adventurous toddlers, and has been dubbed *Half Million Dollar Baby*. Prices for first-rate air racers have, indeed, skyrocketed. *(Reno, 1995)*

The Whittington racer's engine undergoes a compression check as part of its preparation for a race. *(Reno, 1995)*

In 1995, the Phoenix air race organisers divided the Unlimited Class into two subsections: Pro and Vintage Stock. Quite simply, the aircraft in the Pro division were the highly modified racers; while those in the Vintage Stock division were warbirds essentially unchanged from when they rolled off the production line. It was felt that this arrangement would group together the more evenly matched racers, making for more competitive and interesting races. This Sea Fury, *Incredible Universe*, specially painted for the Phoenix 500, was flown to first place in the Stock Gold race by Nelson Ezell. *(Phoenix, 1995)*

The Hawker Sea Fury and Grumman Bearcat are of the same vintage, and were the last of the radial-engined fighters; jets were then on the horizon. As such, these late Second World War fighters were arguably the fastest piston-powered production aeroplanes ever built, and it is, therefore, not surprising that they are prominent on today's race scene. The 'HP' marking on the fin of this Sea Fury stands for Howard Pardue, the aircraft's owner. *(Reno, 1995)*

Brian Sanders of Hesperia, California, flies this plain but powerful Sea Fury. The production line versions of this extraordinary naval fighter, developed at the end of the Second World World War, had a Bristol Centaurus 18 engine using ingenious sleeve valves. Many of the Sea Furies that are raced have been re-engined with American Pratt & Whitney or Wright engines because of the greater possibilities for enhancing their power. *(Reno, 1994)*

Few aeroplanes evoke a sense of raw power as well as the massive Hawker Sea Fury. This example belongs to Sanders Aircraft. *(Reno, 1994)*

Perhaps the most widely recognised Sea Fury is past Gold race winner *Dreadnought*. An assiduously maintained racer, No. 8 commands attention from all directions. *(Reno, 1994 & Phoenix, 1995)*

That gleam of the sun's rays off the aircraft's skin is achieved the old-fashioned way – lots of elbow grease. *(Reno, 1995)*

Tom Camp's *Mr. Awful* is a former Soviet Yakovlev Yak-11. A short-winged, big-nosed bundle of energy, this aeroplane is in a way reminiscent of the old Gee Bee racers. *(Reno, 1995)*

P
14
OWARD PARDUE
F8F-1
MARINES
NL14HP
14

Breckenridge, Texas, is a centre of warbird activities, thanks to the enthusiastic efforts of local resident Howard Pardue, a regular performer on the airshow circuit. The air races could hardly proceed without Howard's participation. His Marine Corps-blue Bearcat, which in representative understatement he calls *Bearcat*, is a regular competitor at the races. Typical of his crew's jocularity is the sitting of an oversize teddy bear on the vintage Harley-Davidson or in the aeroplane's cockpit. With Howard at the controls, this powerful aircraft came in second in the 1995 Reno Silver race at a speed of 361.134 mph. *(Phoenix, 1995 & Reno, 1994 & Reno, 1995)*

Piloted by Bob Hannah of Caldwell, Idaho, this Republic P-47 Thunderbolt competed in the Vintage Stock races at Phoenix in 1995. *(Phoenix, 1995)*

The red, white, and blue *Miss America*, with Museum of Flying pilot Alan Preston at the controls, won the Unlimited Gold race at Reno in 1994 with a speed of 413.773 mph. That year, the fastest contenders competed in a special Super Gold Shootout in which the top speed exceeded 443 mph. *(Reno, 1994)*

First in the pack behind Don Whittington's Griffon-powered P-51 in the Silver race at Reno in 1995 was Howard Pardue in *Bearcat*. Howard was clocked at 361.134 mph. *(Reno, 1995)*

During the early weekday heats at Reno in 1995, uncharacteristic clouds floated over the airport to give the contests an unusual backdrop for a few hours. Oblivious to the overcast, *Bearcat* comes screaming around the pylons. *(Reno, 1995)*

The best-known Grumman Bearcat flying today, Lyle Shelton's *Rare Bear* is flown by superb racing pilot John Penney. *Rare Bear* won the Unlimited Gold Shootout at Reno in 1994 with a speed of 443.296 mph. The following year started off with a win of the Unlimited Pro Gold at Phoenix at a speed of 443.372 mph. Next came a surprise upset second-place finish at Reno with a speed of 465.159 mph. *Rare Bear* flashed across the finish line only about two seconds behind *Strega*. This Bearcat's modifications include the installation of a Wright R-3350 in place of the standard Pratt & Whitney R-2800. Whereas the maximum rating of a production aeroplane was 2,400 hp, in *Rare Bear* it exceeds 4,000 hp. Not surprisingly, fuel consumption has doubled. The propeller came from a Lockheed P-3 Orion, the huge four-engined anti-submarine aircraft. It is no wonder that this aircraft holds the speed record for a piston-powered aeroplane over a 3 km course, set on 21 August 1989 at Las Vegas, New Mexico, with a speed of 528.329 mph.
(Reno, 1994 & Reno, 1995)

John Crocker, flying *Section Eight*, came in sixth in Reno's 1994 Silver race at a speed of 295.940 mph. There is hardly a sound finer than the hum of a Merlin engine.
(Reno, 1994)

A beautifully restored P-51, Robert Converse's *Huntress III* finished third in the Bronze race at Reno in 1994. *(Reno, 1994)*

Ironically, the two-tone tan camouflage scheme makes *Cottonmouth* more easily recognised on the race circuit. Airline pilot Bramwell Arnold of Lemoore, California, flew this old fighter at Phoenix in 1995. He finished in third place in the Vintage Stock Gold race. *(Phoenix, 1995)*

John Herlehy from Mt. Shasta, California, banks hard left. He came in seventh in the Silver race at Reno in 1994. *(Reno, 1994)*

Glistening under the boiling Arizona sun, Jimmy Leeward's *Cloud Dancer* roars around the pylons. The desert's surface, warmed by the constant sun, throws heat into the air, causing turbulence close to the ground. Experienced racing pilots are accustomed to the bumpy ride as they skim over the mesquite. As a safety measure, race rules require that wingtips not dip below the pylons, but in their fiercely competitive drive, pilots sometimes press the margins. Here, *Cloud Dancer* whizzes past a Phoenix 500 pylon. At Reno in 1995, Jimmy finished fifth in the Silver race with a speed of 345.806 mph. *(Phoenix, 1995)*

Sporting a lightning bolt along its fuselage, *Lady Jo* competed in the 1995 Reno Silver race. Rob Patterson of Corona, California, flew the Mustang into eighth place.
(Reno, 1995)

Now known as *Voodoo Chile*, this P-51 was called *Pegasus* when owned by William A. 'Bill' Speer, Jr. A noted aircraft restorer, Bill perished when his other Mustang, *Deja Vu*, crashed during a qualification run at Reno on 12 September 1994. Reportedly, the aeroplane suffered a failure of the propeller governor, and oil splattered over the windscreen and canopy, blinding Bill. Only 48 years old, he had before him at least a few decades of contributions to air racing and aviation in general.

At the 1995 Reno air races, a 'missing man' formation was flown in tribute to those who had been involved in air racing but are now no longer with us. It was a poignant reminder of the special people who have given so much to the world of high-performance flight. Though they have 'gone West,' they are not forgotten. Indeed, Bill's legacy lives on with each pass of racer No. 55. With Bob Hannah at the controls it took fifth place in the Gold race at Reno in 1995, racking up an impressive speed of over 400 mph. *(Reno, 1995)*

Always an Unlimited Gold contender, *Dago Red* zoomed across the sky, finishing third at Reno in 1995. David Price of Santa Monica, California, pushed the heavily modified P-51 to a speed of 449.137 mph. *(Reno, 1995)*

Flying the silver Sea Fury, Dennis Sanders came in sixth in the regular Unlimited Gold race at Reno in 1994. The big aeroplane takes up a lot of sky as it weaves around the pylons. *(Reno, 1994)*

On 6 July 1995, air racing design prodigy Bruce J. Boland passed away. He was 57 years old. Holding degrees in aeronautical and aerospace engineering, Bruce spent his professional career at the super-high-technology Lockheed 'Skunk Works', helping to produce such revolutionary aircraft as the Mach 3-plus Blackbird spyplane and the forbidding Stealth fighter. With the resumption of national air racing in 1964, Bruce volunteered his time and expertise to many of the Unlimited racing teams. He had a special interest in these aircraft, and his contributions helped to expand their performance envelopes.

At the time of his death he was helping to develop two new Unlimited racing aeroplanes, Darryl Greenamyer's *Shock Wave* and the Cornell-Jackson racer. These aircraft, which are reported to be radical hybrids, add to an already brilliant legacy. His burial service included a 'missing man' formation flown by air racing pilots, including Dennis Sanders, the pilot of *Dreadnought*, one of the racers Bruce helped to go faster. The ultimate racing Sea Fury, *Dreadnought*, finished second in the 1995 Phoenix Unlimited Pro Gold race with Dennis Sanders at the controls.
(Phoenix, 1995)

While he is best known for his Bearcat because of his frequent airshow performances in it, Howard Pardue also races a Sea Fury, simply named *Fury*. Faster than his Bearcat, this aircraft came in seventh in Reno's 1995 Gold race at a speed of 389.131 mph. *(Reno, 1995)*

Nelson Ezell was the first winner of the Phoenix Unlimited Vintage Stock Gold race, flying the Sea Fury *Incredible Universe*. His speed was 322.519 mph. *(Phoenix, 1995)*

Regrettably, the sole P-38 Lightning left on the racing circuit, this rare Second World War fighter is one of only about half a dozen still flying in the world. When introduced to the Army Air Forces this aeroplane was considered revolutionary. Sprouting from the fertile mind of Lockheed's Clarence 'Kelly' Johnson, the P-38 had a twin-boom configuration. The propellers counter-rotated to eliminate torque (the tendency of an aeroplane's nose to veer in opposing reaction to the propeller's rotation) and thereby provide a stable platform for the guns mounted in the nose. Powered by two Allison V-12s of 1,425 hp each, the P-38 is quite large and therefore is not noted as a premier racing aircraft. Also, unlike the other fighters, the Lightning has a control yoke instead of a control stick, and Lefty's left hand can be seen gripping the wheel. *White Lightnin'* scored high at Phoenix in 1995, taking second place in the Vintage Stock Gold race. *(Phoenix, 1995)*

Hardly a contender in the Unlimited Class, the Curtiss P-40 was virtually obsolete at the outset of the Second World War. American pilots, notably those comprising the famous volunteer group known as the Flying Tigers, made the most of this old warhorse and tallied up lopsided kill/loss ratios based on superior tactics, honed flying skills, and pure grit. This N model, called *Spud Lag*, was seen bringing up the rear at Reno in 1995. *(Reno, 1995)*

A more recent attempt to achieve the benefits of neutralising torque is the Whittington racer, a radically modified P-51. Fitted with a Rolls-Royce Griffon engine driving contrarotating propellers, this formidable racer won the Pro Silver race at Phoenix in 1995 with a speed of 381.745 mph, and the Silver race at Reno in 1995 at a speed of 379.491 mph. Don Whittington decided to vacate his Silver victory at Reno in order to compete in the Gold race. Although he came in sixth in that race, he increased his speed to 390.456 mph. Air racing needs more innovation, especially in the Unlimited Class, or the sport risks becoming a stale yearly rehash. The stop action picture highlights the contra-rotating propellers. *(Reno, 1995 & Phoenix, 1995)*

Tom Camp's Yakovlev Yak-11 *Mr. Awful* is among the more colourful Unlimited entries. When the Iron Curtain came tumbling down, previously unobtainable Eastern Bloc aeroplanes came on to the market, and the Yak-11, a post-Second World War Russian advanced trainer, was cheap and adaptable. A few, like this one, have been modified with larger engines and are now competing on the air racing circuit. *(Reno, 1995)*

Not a real contender against fighters, this North American T-28B Trojan, has nevertheless, flown in the Unlimited heats at Reno. It belongs to Neil Anderson, an executive with Lockheed Martin at Fort Worth, Texas, who has a very impressive flying background in the military and with the company as a jet fighter test pilot. It must be like travelling in slow motion to go from the Mach 2 F-16 to the radial-engined T-28, but all aeroplanes demand respect. *(Reno, 1994)*

For a sense of sheer power, nothing beats the imposing, radically modified Sea Fury named *Critical Mass*. Fifth in the regular Unlimited Gold race at Reno in 1994 at a speed of 392.134 mph, this racer seems to have genuinely unlimited potential. *(Reno, 1994)*

The Museum of Flying at Santa Monica, California, is keeping a part of aviation history alive by maintaining a growing number of Second World War warbirds in flying condition and even racing them when opportunities permit. This Republic P-47 Thunderbolt raced in the Vintage Stock Class at Phoenix in 1995.
(Phoenix, 1995)

The Museum of Flying's spectacularly restored Supermarine Spitfire Mk XIV participated in the 1995 Phoenix Vintage Stock races. It is a very rare sight on the air racing circuit. *(Phoenix, 1995)*

At another place and another time this scene could be fighters on a sortie, perhaps on a combat sweep over the French countryside. Of course, this was a slice of the action in the Vintage Stock races at Phoenix in 1995. *Million Dollar Baby*, the Mustang in the background, was piloted by Art Vance, Jr. to third place. *(Phoenix, 1995)*

Cloud Dancer flown by Jimmy Leeward of Ocala, Florida, and *Section Eight*, flown by Joe Cenarrusa of Boise, Idaho, battle for position; a contest between Mustangs.
(Reno, 1995)

In a sense, Howard Pardue competed against himself, with both his Bearcat and Sea Fury racing in this heat at Phoenix in 1995. There was no slackening off. *(Phoenix, 1995)*

Airline pilot Stu Eberhardt's Mustang *Merlin's Magic* and fellow airline pilot Lloyd Hamilton's Sea Fury *Baby Gorilla* duel it out at Reno in 1994. *(Reno, 1994)*

A Curtiss P-40E, named *Sneak Attack*, displays on its nose the sharktooth artwork made famous by the Flying Tigers. Only temporarily in the lead, the comparatively slow Warhawk was soon beaten by the other Unlimited racers. *(Reno, 1995)*

In the Silver race at Reno in 1995 Don Crowe of Delta, British Columbia, finished third with a speed of 353.906 mph in his Sea Fury *Simply Magnificent*, just seconds behind Howard Pardue's *Bearcat*.
(Reno, 1995)

Chapter 9
Pilots and Crews

Air race people are as colourful as the paint schemes that adorn their aeroplanes. The spirit of Roscoe Turner lives on. When propellers are not turning, the pilots and crews are busy attending to their aircraft, forever tinkering to squeeze out better performance. Yet they almost always find time to autograph a youngster's programme or answer a fan's question.

Each team puts its stamp on its aeroplane. Usually the personality of the pilot manifests itself in the flight of the racer – for example, high turns and dives along the straightaways or constant low-level pylon hugging. Spectators in the grandstands cheer on their favourites.

Dedicated mechanics toil for uncounted hours in hangars at home and in the pits at the races, preparing their aircraft for a few minutes of furious flight. These often overlooked technicians, attired in their oil-stained overalls, are most appropriately brought into the Winner's Circle to share the limelight when their efforts contribute to triumphant results.

Aglow in the Winner's Circle at Reno in 1995 is *Dago Red*. The third-place finisher with a speed of 449.137 mph, this P-51 from the Museum of Flying is among the crowd's favourites. *(Reno, 1995)*

The fastest piston-powered aeroplane in history, but only the second place finisher at Reno in 1995, *Rare Bear* was towed back to its parking area to try again next year. It crossed the finish line just a couple of seconds after the winner, *Strega*. (Reno, 1995)

Conspicuously streamlined, *Strega* soaks up the glory in the Winner's Circle at Reno in 1995. Its winning speed of 467.029 mph was the third fastest for a Reno Gold race champion. *(Reno, 1995)*

Bill 'Tiger' Destefani, still attired in his fire-retardant flight suit, accepts the National Championship Air Races Trophy for his victory in the Unlimited Gold race at Reno in 1995. Race officials and an announcer from the ESPN sports channel made the presentation shortly after the racers landed. *(Reno, 1995)*

David Price and the *Dago Red* crew pose for a picture with their vibrant aeroplane. A third-place finish in such a competitive field is sufficient reason to be all smiles. *(Reno, 1995)*

Rare Bear has certainly enjoyed its share of the limelight. As Gold race champion at Reno in 1994, its pilot, John Penney, was swarmed by media representatives when he climbed down from the cockpit after his win. *(Reno, 1994)*

Taking the second-place finish in his stride was *Rare Bear* owner Lyle Shelton. Happily signing autographs on air racing programmes for well-wishers, he could be overheard saying, with a smile on his face, that his aeroplane's engine needed a little tweaking for the races next year. *(Reno, 1995)*

A jubilant Tiger Destefani and his proud crew give the aviator's sign of affirmation. The closeness of the two top competitors at the finish line made the win so much more glorious for the *Strega* team. *(Reno, 1995)*

There just would not be air racing without the legendary Robert R. A. 'Bob' Hoover in his green-striped flying suit and straw sombrero. Like the investment firm that used to say everyone listened when they spoke, when the godfather of air racing speaks the racing pilots and the media pay rapt attention. A matter of weeks after this picture was taken at Reno in 1995, Bob's US medical certificate was reinstated. It had been withdrawn about 2½ years earlier under extremely controversial circumstances. During the hiatus, Bob obtained a medical certificate from Australian authorities and kept on flying in countries that recognised that certificate. Note the Australian shoulder patch on his flying suit. *(Reno, 1995)*

Pilots seem unable to converse without using their hands. Even the best, like David Price, lack immunity from this affliction. Charged up by a super performance only minutes before in *Dago Red*, David is brimming with excitement. *(Reno, 1995)*

Delmar and Linda Benjamin have brought countless thrills to airshow audiences across America. The Gee Bee R-2 replica that Delmar flies is an excellent example of the 1930s original that rolled out of the Granville brothers' shop. Delmar's aerobatic prowess is superlative. Not surprisingly, fans clamber for his autograph, even in the absence of paper.
(Reno, 1995)

Although they never seem to garner as much attention as their Unlimited racing cohorts, the pilots and crews of the slower racers are equally dedicated and hardworking. When the last propeller stops turning, these men and women of the air gather in the pit area to collect their thoughts and enjoy good company. Ideas for improving next year's performance are mulled over. Alfred Goss, past AT-6 Class champion and owner of *Warlock*, takes a moment to stand with his teammates after the races. He is fourth from the left in the second row. *(Reno, 1995)*

Veteran racing pilot Alan Preston of Pacific Palisades, California, explains race manoeuvrings, also with his hands. *(Reno, 1995)*

While he was not the winner at Reno in 1995, Al Goss is nevertheless elated to be doing what he loves best. *(Reno, 1995)*

Chapter 10
Nose Art Gallery

Racing aeroplanes would hardly fit the bill were it not for the decorative and clever artwork emblazoned on them. Walking the ramp at Reno and Phoenix is rather like strolling through an art gallery, for in addition to the sparklingly maintained aircraft, which are arguably works of art in themselves, almost every one features a painted illustration on its nose. Some of this nose art is funny, some is serious, and some is historical. It is all interesting.

Pilots have a way of capturing their outlook on life or reflecting their personality in a word or a phrase. Such is the case with aircraft nose art; a kind of public statement by an aeroplane's owner. Usually decorative, nose art is incisive as well. *Warlock*, is the name of racer No. 75, Al Goss' past AT-6 Class winner. *(Reno, 1995)*

Perhaps the most appropriately nicknamed P-51 on the ramp at Reno is racer No. 22. The Rolls-Royce Merlin engine, built under licence by both Allison and Packard, has a distinctive hum that is music, or magic, to the observer. *(Reno, 1995)*

Airline pilot Sherman Smoot of Templeton, California, flew racer No. 86 *Bad Company* to second place in the 1995 Reno AT-6 Gold race, a mere fraction of a second behind the champion, Charles Hutchins, in *Texas Red*. *(Reno, 1995)*

Racer No. 37, *Tinker Toy*, has been a part of AT-6 Class racing for many years. *(Reno, 1994)*

As awesome as the Sea Fury itself is the title *Dreadnought*. *(Phoenix, 1995)*

Emblazoned upon a most patriotic backdrop is the name *Miss America*. This artwork decorates the P-51 of Oklahoma City neurosurgeon Brent Hisey. *(Phoenix, 1995)*

Another appropriate name, *Risky Business*, appears on the nose of Bill Rheinschild's Mustang, racer No. 45. *(Reno, 1995)*

Nose art on a Mustang that says it all: *The Believer*. *(Reno, 1994)*

Emblazoned just below the exhaust stacks of racer No. 68 is the name *Sparky*, for obvious reasons. For good measure, several sparks are painted on the nose of the Mustang. *(Reno, 1994)*

Cropdusting pilot Nick Macy of Tulelake, California, came in fifth in the 1994 Reno AT-6 Gold race flying racer No. 6, *Six Cat*.
(Reno, 1994)

A combat motif like those employed on Mustangs during the Second World War is seen on racer No. 27, *Section Eight*.
(Phoenix, 1995)

The author/photographer, Philip Handleman, his image reflected in the polished aluminium of a restored aircraft, focuses on his subject at an air race. *(Phoenix, 1995)*